THEORY OF CONSCIOUSNESS

ALAN J. BENESI

ARPress

ILLUMINATING IDEAS
EMPOWERING VOICES

ARPress LLC
45 Dan Road Suite 5
Canton MA 02021
Hotline: 1(888) 821-0229
Fax: 1(508) 545-7580

Ordering Information:
Quantity sales. Special discounts are available on quantity purchases by corporations, associations, and others. For details, contact the publisher at the address above.

Printed in the United States of America.

ISBN-13:	Softcover	979-8-89330-017-8
	Hardcover	979-8-89330-018-5
	eBook	979-8-89330-019-2

Library of Congress Control Number: 2024900799

CONTENTS

Chapter 7

Chapter 8

Chapter 9

Chapter 10

Chapter 11

Chapter 12

The Collected Postulates
of Consciousness

PREFACE

My interest in consciousness was stimulated by hallucinogenic drugs. Their effects were, for me, intense and beautiful. I wondered how simple chemicals were able to open the doors of human perception[1]. After a small dose of LSD on a warm Spring afternoon in 1970 I had a revelation. Tiny flies swarmed in the sunbeams hitting the trunk of a huge redwood. Then without warning the truth sent a jolt of realization through my being and I saw the humming life in all things. It was obvious. Everything is conscious. It was a sensory experience, not a result of logic or language. But it also made everything fit logically! This epiphany is the origin of my continuing interest in consciousness.

That everything is conscious is not a new idea.[2] I do believe, however, that I have recognized a simple and scientifically rigorous way to look at the phenomenon of consciousness. I earned a Ph.D. in Biophysics at UC Berkeley, and I spent my academic career

---

1   Aldous Huxley, The Doors of Perception, Harper & Row; Clean & Tight Contents edition (1970)

2   see Greek Hylozoism, Panpsychism, Hindu Pantheism

teaching Chemistry and doing Nuclear Magnetic Resonance (NMR) Spectroscopy at major universities. I am familiar with physics, quantum mechanics, chemistry, biophysics, and biochemistry.[3]

I've attempted to be scientifically rigorous. I use axioms and postulates to present the main ideas. The axioms are assumed to be self-evident, i.e., inherently true. The postulates are consistent with observations so far. I try to adhere to the principle, often called Occam's Razor, that the simplest idea that conforms with observations is the best. The ideas will be controversial.

---

3   But I readily acknowledge that there is still much about science and consciousness that I don't know or haven't mastered

Chapter 1

DEFINITIONS OF CONSCIOUSNESS AND THE QUEST FOR MEANING

The definitions of consciousness found in dictionaries are vague. They illustrate the lack of clarity surrounding the subject. Three are presented below:

1'st set of definitions:[4]

1. *a* : the quality or state of being aware especially of something within oneself

   b : the state or fact of being conscious of an external object, state, or fact

   c : awareness; *especially* : concern for some social or political cause < *The organization aims to raise the political consciousness of teenagers.* >

2. the state of being characterized by sensation, emotion, volition, and thought : mind

3. the totality of conscious states of an individual

---

4   https://www.merriam-webster.com/dictionary/consciousness

4. the normal state of conscious life <*regained consciousness*>

5. the upper level of mental life of which the person is aware as contrasted with unconscious processes

2'd set of definitions:[5]

1. the state of being conscious; awareness of one's own existence, sensations, thoughts, surroundings, etc.

2. the thoughts and feelings, collectively, of an individual or of an aggregate of people:

the moral consciousness of a nation.

3. full activity of the mind and senses, as in waking life:

to regain consciousness after fainting.

4. awareness of something for what it is; internal knowledge:

consciousness of wrongdoing.

5. concern, interest, or acute awareness:

class consciousness.

6. the mental activity of which a person is aware as contrasted with unconscious mental processes.

7. Philosophy. the mind or the mental faculties as characterized by thought, feelings, and volition.

---

5   http://www.dictionary.com/browse/consciousness

8. Idioms: raise one's consciousness, to increase one's awareness and understanding of one's own needs, behavior, attitudes, etc., especially as a member of a particular social or political group.

Third set of definitions:[6]

Sentience, awareness, subjectivity, the ability to experience or to feel, wakefulness, having a sense of selfhood or soul, the fact that there is something "that it is like" to "have" or "be" it, and the executive control system of the mind, or the state or quality of awareness, or, of being aware of an external object or something within oneself.

Another way to understand consciousness, one closer to the ideas presented here, is the famous quotation:[7]

"I think, therefore I am."

Even Descartes' statements presume human consciousness, but how far down the evolutionary tree does one have to go before consciousness "disappears"? What about your dog, your cat, a frog, a fungus, a plant, a single celled organism, a virus, a molecule, an atom? Do these all possess consciousness? What has consciousness? And more fundamentally, and from a number of perspectives, what is consciousness? What are its properties? How does it work? What can it do? These are some of the fundamental questions discussed below.

---

6   https://en.wikipedia.org/wiki/Consciousness

7   Rene Descartes (translated by Donald A. Cress), Discourse on Method, 3'd edition, Hackett Publishing Company, 1998, ISBN: 0872204227

Chapter 2

THE ENTITY PERSPECTIVE VS. MECHANISTIC PERSPECTIVES

lassical physics tells us that an object experiences changes in momentum only if a force acts on it. Specifically, **F(t)** = m **a(t)**, where **F(t)** is the force vector at time t, m is the mass of the object, and **a(t)** is the acceleration vector at time t of the object being acted on by the force. The following analysis is based on that principle.

Now consider a human being, and more specifically consider yourself sitting on a chair watching a show on TV. It's a warm day, and your thoughts (i.e. your consciousness) keep(s) turning to an ice cold beer. An advertisement comes on, and you decide that it's time to go to the refrigerator and get the cold beer you are thinking about. You stand up and walk to the refrigerator and get the cold beer, then return to your chair so you can see the rest of the TV show. From *your entity perspective*, your consciousness was the instigator of this behavior. **You know exactly WHY** this behavior occurred. **_You made it happen!_**

Let's now analyze the same sequence of events using the **_mechanistic macroscopic perspective_** of classical physics without presuming consciousness of the "object", i.e. without the entity perspective. There is a time dependent acceleration **a(t)** of the center of mass of the object as it goes to the refrigerator and then returns. In the absence of the entity perspective, these accelerations must be instigated by a time dependent force **F(t)** on the object. We can examine the forces exerted by your legs and other body parts, but we have *no clue* *WHY* the forces occurred.

Next, we consider the ***mechanistic microscopic perspective*** from physics, biophysics, chemistry, biochemistry, and neurochemistry. At the molecular level and smaller, the analysis requires quantum physics. These perspectives also presume that the material universe is constructed of objects subject to forces although the objects are much smaller and have both wavelike and particle-like properties. There is no consideration given to the entity perspective. From this perspective a human brain is constructed of neurons and support cells which are in turn constructed of biological molecules. Between neurons there is synaptic transmission involving release and binding of neurotransmitter molecules. Along neuronal axons there are action potentials involving influx of external sodium ions and efflux of potassium ions proceeding as "electric" waves. At neuromuscular junctions similar processes occur, with calcium ions most obviously involved. From this perspective the change in momentum of the human entity is nothing but the causal result of the sequence of neurochemical events in the brain, and the changes in momentum are causal results of muscle contractions stimulated by the causal events in the nervous system. The whole process is as mechanical as billiard balls colliding and follows the laws of physics and chemistry exactly. The biological structures are amazing and seem to be constructed intelligently, but we still have **no clue WHY** these microscopic forces exist.

Chapter 3

ENTITIES

TO UNDERSTAND CONSCIOUSNESS, WE START WITH THE ASSERTION:

Axiom 1: The entity perspective is valid.

_ou_ really exist. Your mind and your consciousness really exist. You can change your momentum. You can decide not to change your momentum.

With this axiom we recognize that entities exist and that they possess consciousness. Within constraints, the consciousness of an entity controls its behavior. I think, therefore I am. My dog thinks, therefore, he is. The fundamental failing of the mechanistic perspective in either the macroscopic or microscopic forms is that it does not explain the existence of consciousness. It does not recognize the role of consciousness in instigating "forces" that affect the motions of "objects". Consciousness *must* be incorporated into the scientific perspective.

From the *entity perspective*, consciousness is the instigator of time dependent forces. Therefore, we assert:

Axiom 2: An entity can use its consciousness to change its own momentum, i.e. "create" a force on itself.

This is probably the best definition of an entity.

Definition 1: An entity is a consciousness that can cause changes in its own momentum, i.e. forces.

A human entity can also use a bulldozer to move boulders. A dog entity can herd a flock of sheep. A pride of lion entities can bring down a water buffalo. It is apparent that entities can change the momentum of other "objects" and entities. Therefore, we assert:

Axiom 3: An entity can cause changes in momentum of "objects" and entities other than itself.[8]

It is self evident that humans and dogs possess consciousness, intelligence, and feelings. Which entities do not possess these properties? It is proposed here that there is no reason to doubt that all entities have similar qualities. Therefore, we postulate:

Postulate 1: Entities possesses intelligence that affects their behavior.

Postulate 2: Entities possess feelings that affect their behavior.

---

8   Conservation of momentum requires this even in the absence of consciousness, but only the entity perspective answers ***WHY these*** and not other momentum changes occurred.

CONSCIOUSNESS MAKES THINGS HAPPEN: FORCE AND CONSCIOUSNESS, A UNIFIED THEORY OF PHYSICS

Because it makes sense and is consistent with the entity perspective, we postulate:[9]

Postulate 3: *All* changes in momentum, i.e. forces, are caused by intelligent behavior of conscious entities.

And it's corollary postulate:

Postulate 4: The presence of forces indicates the presence of intelligent conscious entities.

Since forces are present within and between all forms of matter and energy, from subnuclear particles to the Universe itself, there are entities at work in all these realms. We therefore postulate:

Postulate 5: Intelligent entities exist in different space-time realms and thereby produce "forces" in the realms. These include the nuclear force-nucleus realm, the electromagnetic force-atom/molecule realm, the gravitational force-astronomical realm, and the biological force-biological entity realm.

Postulate 6: All matter and energy is inhabited by intelligent conscious entities.

---

9   This is the Most Important Postulate

Postulate 7: All entities in all realms are conscious simultaneously.

These postulates are consistent with observations, as shown in the next chapters.

Chapter 4

IDENTIFICATION OF ENTITIES

Establishing material and physical boundaries between entities and their surroundings is difficult. Where does one entity "start" and where do adjacent entities, i.e. the surroundings, "begin"? To clarify this issue, we postulate:

Postulate 8: The "boundary" of an entity is defined by the region of space-time within its conscious observation and control.

Delineation of the boundaries depends on the details. For example, a human entity has receptors for sound, light, taste, odor, pressure, heat, pain, etc. The boundaries for each sense are different. The consciousness of the human entity has control of its body, muscles, and behavior. Most human entities exhibit behaviors over relatively small space-time regions on the surface of the Earth. But human entities have access to information about stars, planets, and galaxies. This extends the boundary for conscious observation in both space and time.[10] Sending a spacecraft to land on and explore Mars extends the

---

10   Astronomers on Earth see the light emitted by distant galaxies billions of years ago!

boundary of human conscious control. At the time of this writing, humans have extended the "reach" of their conscious control to the outer parts of our solar system.

Chapter 5

SPACE-TIME REALMS OF ENTITIES

Because we have postulated that all forces are created by conscious entities, we recognize that different entities operate in different space-time realms. For example, strong and weak nuclear forces hold the nucleus together and cause radioactive decay, respectively, but their range is confined within the nucleus. The *longest* time scale T_{nuc} pertinent to nuclear forces is obtained by dividing the nucleus diameter $d_{nuc} \sim 1 \times 10^{-15}$ m by the speed of light $c = 2.998 \times 10^8$ m s^{-1}, i.e. $T_{nuc} = d_{nuc}/c \sim 3.3 \times 10^{-24}$ sec. Electromagnetic forces totally dominate the atomic and molecular space-time realm (with $d_{mol} \sim 1 \times 10^{-10}$ m), so $T_{mol} = d_{mol}/c \sim 3.33 \times 10^{-19}$ sec. Unlike nuclear forces, electromagnetic forces have infinite range and therefore also exert effects in larger-longer space-time realms. Gravitational forces also have infinite range and dominate the astronomical extremely-large/extremely-long space-time realm with, for example, $T_{galaxy} = d_{galaxy}/c \sim 100,000$ to 1 million years! [11] These are examples of the space-time realms of the four fundamental types of elementary force that are presently known.

---

11   In black holes and neutron stars, gravitational forces are stronger than subatomic and nuclear forces.

Chapter 6

EVOLUTION OF BIOLOGICAL ENTITIES

The axioms and postulates presented so far are consistent with the theory of biological evolution. The evolution of entities has been from small to large. After the big bang came the elementary particles: quarks, electrons (fermions), gluons and photons (bosons), and atomic nuclei. Gravity created stars, galaxies, and solar systems. Atoms and molecules formed, then in lucky places with liquid water biological macromolecules, viruses, algae, bacteria, single celled organisms, plants, multicellular organisms, …, dogs, humans. Where does consciousness first appear in this sequence? Where does it disappear in the reverse sequence? We assert that consciousness never "appears" or "disappears". Any dividing lines are arbitrary and based on an anthropocentric point of view.

Biological entities are characterized by an amazingly intricate and optimized chemical machinery of biological molecules, organelles, cells, and multicellular organs. The operation of these structures and molecular machines suggests that they are designed and operated by intelligent entities. Indeed, they are! The entities responsible operate in the quantum realm.[12] The structures that the quantum realm entities build are analogous to blueprints (DNA), energy plants (mitochondria and chloroplasts), machines (muscles, ribosomes), factories, storage facilities, roads, and the internet of the brain: the neurons and support cells, the synapses, the axons, the dendrites.

Overall, we postulate that quantum realm intelligent entities design and operate biological molecules and structures:

Postulate 9: Biological molecules and structures are inhabited and created by quantum realm intelligent entities.

Postulate 10: The behaviors, intentions and consciousnesses of substructural entities create the behaviors, intentions and consciousnesses of the entities they build.

Apparently unlike the quantum realm entities that create them, biological entities share the property that they are based on DNA or RNA "blueprints". Except for viruses, they require input of energy (i.e. food) and produce waste products. They are "dissipative", like eddies in a river.[13] They eat food and excrete waste products. They require this flow for their existence.

Conjecture 1: Perhaps all entities are dissipative from their own perspective.

---

12   If one looks at the atomic level structures of non-biological substances, forces and "intelligent arrangements" are also present.

13   Dilip Kondepudi and Ilya Prigogine, Modern Thermodynamics: From Heat Engines to Dissipative Structures (Coursesmart) 2nd Edition, Wiley, 2014 ISBN-10: 111837181X

Conjecture 2: Perhaps all entities within their respective realms possess a similar organization to that of biological entities...again from their perspective.

Regardless of the details, we propose that biological entities are built and inhabited by entities operating at smaller-faster space-time levels in the quantum realm. These entities create us. They are the gods within us.

This analysis also supports the idea that *in general* intelligent entities operating in smaller-faster space time realms give rise to intelligent entities in larger-slower space-time realms.

Chapter 7

ENTITIES IN THE QUANTUM REALMS

What or who are the intelligent entities in the quantum realm? Are quantum realm "objects" that experience changes in momentum such as electrons, photons, and nuclei individual entities or perhaps entire civilizations or universes of smaller entities? Why are the four known types of elementary force so apparently simple in comparison to the more complicated, irregular and seemingly arbitrary time dependent forces created by biological entities like humans and dogs?

In this chapter we consider forces that affect very small "objects" such as molecules, atoms, electrons, photons, nuclei, quarks, gluons, and other nuclear particles. Successful *mechanistic* description of these systems requires analysis in terms of quantum physics. In its most accurate formalism, the "forces" observed for these "objects" are described using the theories of Quantum Electrodynamics (QED) for electromagnetism, Quantum Chromodynamics (QCD) for the strong nuclear force, and Quantum Flavordynamics (QFD) for the weak nuclear force respectively. The QED prediction, for example, agrees with the experimentally measured value of the electron's magnetic moment to more than 10 significant figures, making it the most accurate prediction in the history of physics.

A bizarre aspect of these quantum realm forces is the mechanism by which they are created. Forces between wave-particles called fermions are mediated by force carrier virtual wave-particles called gauge bosons.[14] Virtual wave-particles do not "escape" and are not directly observable to external observers. They act as momentum transfer particles between the fermions. They are the "agents" of forces in the quantum realm. All gauge bosons have integer spin, i.e. 0, 1, 2, …[15] All elementary fermions, e.g. electrons and quarks, have half spin 1/2. Composite fermions such as nuclei are called "hadrons" and are composed of multiple elementary fermions. Hadrons have spins varying in units of ½. The smallest nuclear spin quantum number is 0, followed by ½, 1, 3/2, 2, 5/2, etc.[16]

In the atomic and molecular realm, (QED realm), electromagnetic forces are mediated by exchange of chargeless, massless, spin 1 virtual photons[17] with energy $E = h\nu$, where E is the energy of the photon, h is Planck's constant, and ν is the frequency of the photon.[18] The

---

14   https://en.wikipedia.org/wiki/List_of_particles

15   https://en.wikipedia.org/wiki/Spin_%28physics%29

16   I am a nuclear magnetic resonance spectroscopist. NMR spectroscopists measure and interpret the radiofrequencies emitted by nuclei (hadrons) placed in magnetic fields for nuclei with spin ½, 1, 3/2, 2, 5/2, 3,7/2,4,9/2, etc. Spin 0 nuclei are not observable with NMR.

17   *Virtual* bosons can't be detected by external observers since they are exchanged, not emitted to the external universe.

18   Some quantum realm processes cause photons to be emitted to the environment. Our eyes see the (human) visible photons.

momentum p of the photon is given by p = hv/c and c is the speed of light.[19] Virtual photons transfer momentum, i.e. act as *force carriers,* between fermions such as electrons and nuclei. They travel at the speed of light. They are, in fact, short lived light wave-particles.

In the nuclear realm, the strong nuclear force (QCD realm) is mediated by exchange of 8 types of massless spin 1 virtual wave-particles with no electric charge called gluons. The gluons transfer momentum between mass-bearing spin ½ fermions called quarks that have an electric charge of either +2/3 or -1/3 (the proton charge is +1, the neutron charge is 0). It takes three quarks of the right types to make a proton or neutron. Almost 99% of the mass of the protons, neutrons, and the nuclei they build arises from the relativistic effects of the near light speed motions of the quarks caused by their interactions with gluons. The nucleus is a seething mass of quarks and gluons moving at nearly the speed of light within the tiny volume of the nucleus. The QCD model of the strong nuclear force is much more complicated than the QED model of the electromagnetic force.[20,21] This is amazing and is a clear indicator of the difference between the nuclear and atomic quantum realms. The mass of a proton is almost 100 times greater than the sum of the rest masses of the three quarks that make it up, and the gluons have zero rest mass. The same is true for the neutron, although the three quarks that make up the neutron are different. The huge relativistic contribution to the mass of nuclei suggests that the *nuclear*

---

19   Alan J. Benesi, The Gods Are Within Us, Kindle Direct Publishing, 2022, ISBN: 9798839922709

20   ibid

21   http://hadron.physics.fsu.edu/~crede/forces.html

quantum realm is smaller and faster than the larger and slower *atomic/ molecular* quantum realm where electromagnetic forces are mediated by photons.

This observation also fits well with the idea that in the ultimate analysis consciousness mediates all changes in momentum, and that "mass" and objects are manifestations dependent on the space-time realm of the observing entity. We therefore postulate just that:

Postulate 11: Ultimately, all that exists are consciousness-mediated changes in momentum. Mass and objects are merely manifestations that depend on the space-time realm of the observing entity.

We can make the connection between mass and energy more explicit by invoking Einstein's famous equation for the energy content of a rest mass m, $E = mc^2$.[22] Equating this to the energy of a photon, $E = h\nu$, yields the rest mass equivalent of a photon, $m = h\nu/(c^2)$.

The weak nuclear force (QFD) carriers are mass-carrying spin 1 bosons called W and Z bosons, -1, +1, and 0 charged respectively. These can change a quark from one "flavor" to another and are responsible for radioactive decay.

Mass is transferred between particles by the Higgs boson, which has spin 0 and no charge but a large mass. Its existence was recently verified experimentally.

---

22   Albert Einstein, The Principle of Relativity, Dover, 1952 ISBN-10: 0486600815

It is bizarre that nature behaves this way. Not all the details have been worked out. It's as if everything was conscious, intelligent, and arbitrary in certain ways! Who'd have set it up this way? Yes, you guessed right, the intelligent entities who live there!

The gravitational force may have its own boson, the graviton (spin 2, chargeless, massless, very low energy, and travels at the speed of light), but no experimental evidence has yet been obtained.

Chapter 8

THE HEISENBERG UNCERTAINTY PRINCIPLE AND THE ENTITY PERSPECTIVE IN QUANTUM SPACE-TIME REALMS

The Heisenberg Uncertainty Principle[23] is the key to understanding space-time realms. It applies to the quantum realms (QED, QCD, and QFD realms) from the human biological entity perspective. One form is:

$$\Delta E \, \Delta t \geq \hbar/2 \quad \text{(Equation 8-1)}$$

where $\hbar$ is Planck's constant divided by 2π, ΔE is the uncertainty in the energy, and Δt is the uncertainty in the time. The minimum allowed for the $\Delta E \, \Delta t$ product is $\hbar/2$.

The uncertainty in time Δt indicates not only the uncertainty in the duration of time but also the uncertainty in the direction of time in the quantum realm. Quantum Electrodynamics, for example, includes contributions from energy transfers that occur both forwards

---

23   W. Heisenberg, "Über den anschaulichen Inhalt der quantentheoretischen Kinematik und Mechanik", Zeitschrift für Physik (in German), **43** (3–4): 172–198, 1927

and backwards in time.[24] This becomes easier to comprehend when one realizes that according to special relativity a photon *experiences* no time in passing from point A to point B or from point B to point A. Bizarre!

Consider an example of an atom in which the electrons and nuclei are bound together by electromagnetic forces, hence by exchange of virtual photons. The uncertainty in the energy of virtual photons causes no change in the total energy of the atom. However, the uncertainty principle predicts that if $\Delta E \approx 0$ the uncertainty in time $\Delta t \approx \infty$. This means that *from the atomic realm perspective* the effects of almost all possible virtual photon exchange processes contribute over very long-time spans, thereby averaging to the relatively simple observed behavior described by quantum mechanics. In other words, the Heisenberg Uncertainty Principle allows for averaging and hence "smoothing" of the behavior.

Because the uncertainty in the energy ΔE cannot be exactly 0, the Heisenberg Uncertainty Principle maintains that time also exists for the entities that occupy the quantum realms, but that biological realm entities can't make measurements on such small-short space-time scales. Nevertheless, for the atomic realm entities themselves time exists just as it does for biological entities.

The same holds true for the strong nuclear force, where the effects of all possible gluon exchanges average to yield the observed behavior

---

24   repulsion in forwards time is attraction in backwards time.

(for example the masses of the proton and neutron); and for the weak nuclear force, where exchange of W and Z bosons causes radioactive decay of nuclear particles.

Therefore, we postulate:

Postulate 12: Intelligent entities in the quantum realm live at almost infinitely faster speed than intelligent entities in larger-slower realms.

Conjecture 3: From the perspective of individual entities that "live" in a space-time realm, their behaviors are intelligent and complicated and time dependent...like behaviors of biological entities.

I believe that time, history, and evolution occur in all space-time realms. Civilizations are created and perfected. The "universe" for the realm is inhabited and eventually becomes controlled by intelligent entities and stabilized in its behavior. Every aspect of the process is incorporated into the average and the steady state behavior. *None* of the details are observable in larger-slower realms.

If the individual entities are like biological entities, their behavior (for example a "map" of their locations in their own space time) looks arbitrary and bizarre. Think of yourself going to and from work with occasional trips to the store, friends, and relatives. Such a map for an individual entity over time would look bizarre and arbitrary, although the behavior is totally understandable from the entity perspective. Following the map for very long times provides some "smoothing" in the behavior but would still look arbitrary.

Now consider the same behavior map for a *statistical ensemble of entities* acting over very long time spans. This would provide "total smoothing" of the map. This seems to be the case for the quantum realms. The regular quantum realm mechanistic behavior observed by biological entities must represent behavior of statistical ensembles of intelligent entities, hence:

Postulate 13: Behavior observed by entities in larger slower realms represents the averaged and smoothed intelligent behavior of statistical ensembles of entities from smaller faster realms.

Conjecture 4: Entities are fractal in nature,[25] with structure and consciousness at smaller-faster space time levels leading to structure and consciousness at larger-slower space-time levels.

If force is the manifestation of intelligent behavior of conscious entities, it follows that force carrier particles, i.e. bosons, are consciousness wave-particles.[26,27]

Postulate 14: Bosons are force carrier wave-particles; hence they are consciousness wave-particles.

The *quantum realm entities* exist on space time scales so small and fast that only their ensemble average behavior (electromagnetic, QED, strong nuclear forces, QCD, weak nuclear forces, QFD) is apparent at larger slower space time scales. Therefore, we postulate:

---

25   Benoit B. Mandelbrot, Benoit B.; The Fractal Geometry of Nature .New York: W. H. Freeman and Co., 1982. ISBN 0-7167-1186-9

26   Let there be light

27   I saw the light

Postulate 15: All nuclei, nuclear particles, electrons, atoms, and molecules, are designed, inhabited, and operated by quantum realm intelligent entities.

"Where" are the entities and what do they look like? Do they know about the consciousnesses they build and the forces they produce? Do they know and build our feelings? Does their averaged behavior reflect utopian civilizations?

Postulate 16: The fact that their averaged behavior is consistent over time suggests that entities at all levels seek stability.

We presently have no way of knowing what the entities are. We cannot and perhaps never will be able to observe the way in which their behavior averages to create forces. But because they build the civilization that is a biological entity we assert:

Postulate 17: Biological entities evolve with time because the designs of the substructural entities are not perfect. Evidently the substructural entities can make mistakes and learn from them. This also confirms that time exists for them.

It is reassuring to know that "no one is perfect", not even in the quantum realm.

Chapter 9

GRAVITATIONAL FORCES AND ASTRONOMICAL ENTITIES

At the time of this writing, gravitational waves of the type predicted by Einstein's Theory of General Relativity[28] have been experimentally observed. The observed waves were generated by collapse of two black holes into one. Up to now we have proposed that the electromagnetic and nuclear forces that govern the behavior of molecules, atoms, nuclei, and nuclear particles are created by quantum realm entities that operate on such small and fast space time scales that only the average of their behavior is observed. But what about gravitational forces that govern the motions and evolution of stars, planets, and galaxies, black holes, and neutron stars? Are these created by entities that operate at small and fast space-time scales or by entities that operate at large and slow space-time scales? *Or both?* Einstein's theory of general relativity ascribes gravitational forces to curvature of space time.[29] Other theories ascribe gravitational forces to exchange of bosons (spin **2,** massless and chargeless) called

---

28   Albert Einstein, The Principle of Relativity, Dover, 1952 ISBN-10: 0486600815

29   ibid

gravitons.[30,31] Like the other massless bosons, they travel at the speed of light. The existence of gravitons (which are very low energy) has not been experimentally confirmed at the time of this writing.

The gravitational space-time realm can be extremely small and fast. It can contribute to the atomic and nuclear space-time realms under special conditions due to the enormous densities and gravitational fields found in neutron stars and black holes.

The gravitational space-time realm can also be extremely large and slow. Gravitational forces are the creators of stars and planets and galaxies and clusters of galaxies.

30   https://en.wikipedia.org/wiki/List_of_particles

31   https://einstein.stanford.edu/content/relativity/q1669.html

Chapter 10

TIDINGS OF COMFORT AND JOY! HUMAN DEATH IS NOT THE END OF CONSCIOUSNESS. THE GODS ARE WITHIN US

The entity view of consciousness has religious implications. From the entity perspective human consciousness disintegrates along with the amazing biomolecular machines and processes that create it. However, we can rejoice in the fact that the quantum realm conscious entities that create us survive our deaths and know every detail of our lives and our experiences. They design our behavior and build the molecular machines that make us work. It is my belief that when humans die, their consciousness disintegrates into the intelligent entities that exist in the quantum realm, where consciousness and feelings also exist.

Postulate 18: Human death is not the end of consciousness. It is a return to the quantum realm of the entities that make us. Individual humans die, but consciousness, intelligence, and feelings persist in quantum realm entities after human death.

Consciousness exists in the form of intelligent entities at all space-time levels. Small-fast entities evolve and form stable "objects" and "forces" first. They create larger-slower entities, who in turn create even larger-slower entities. Since the big bang (a subject for further consideration), evolution has been "bottom up". The gods are within us.

Chapter 11

FORCES CREATED BY BIOLOGICAL ENTITIES

We have discussed the four fundamental types of force: electromagnetic, strong nuclear, weak nuclear, and gravitational. We have proposed that the elementary forces operating in the quantum realm are created by conscious entities that operate at smaller-faster space-time scales than biological entities such as humans and dogs, and that gravitational forces operate in both the astronomical realm at a large-slow space-time scale and in the quantum realm at a small-fast time scale. Here we focus on the erratic, intelligent, and specific "forces" and behaviors created by biological entities such as humans and dogs. Biological entities are not the creators of a fundamental force, at least from their own "in-realm" perspectives. They necessarily operate on a slower space time scale than the quantum realm entities that create them. Conversely, they operate on a much smaller and faster time scale than astronomical entities.

Therefore, we ask, does the erratic behavior exhibited by biological entities "average" to yield reproducible and relatively simple forces in the larger slower astronomical realm? And what would it mean to take the space-time average over "all time"? What role do biological entities play in the evolution of the astronomical superstructure? We can't answer these questions, but instead offer the following conjecture:

Conjecture 5: The behavior of biological entities determines long term astronomical realm behavior, i.e. the long term behavior of our universe.

If this conjecture is right, it is our destiny to colonize and contribute to the organization of the astronomical realm.

Chapter 12

STREAM OF CONSCIOUSNESS

D o I claim that the postulates presented in this theory are absolutely correct? No. I can only claim that they are consistent with observations, hence legitimate postulates. They provide a starting point for further analysis. I don't know if there is a way for human entities to get around the limitation of the Heisenberg Uncertainty Principle. If it holds true, human entities will never know the details of "life in the quantum realm". Similarly, we are apparently incapable of experiencing "life in the astronomical realm" because human entities live and die before major changes occur. But they can at least use telescopes to see ancient major astronomical changes due to the vastness of the Universe. Some bosons emitted billions of years ago are only now being detected, because they are limited to traveling at the speed of light.

Occam's Razor[32] suggests that the entity perspective is correct. The idea of equating force and conscious behavior is simple and beautiful. The idea of intelligent beings living simultaneously in different space-time realms is mysterious and explains everything. It is

---

32   the simplest model that is consistent with observations is the best.

bizarre but logical that statistical ensembles of intelligent entities that live in smaller faster realms design intelligent entities that live in larger slower realms.

If this view is correct, we are all entities living simultaneously in our respective realms. We have recognized the existence of nuclear, atomic, biological, and astronomical realms. We have concluded that larger-slower realms are built from smaller-faster realms and by smaller-faster intelligent entities. Individual entities in different realms do not seem to be able to communicate directly with each other, although they do communicate indirectly, in that the smaller-faster realm intelligent behavior averages to yield "object-like" behavior in larger-slower realms. Communication between realms seems to be achieved via bosons traveling at the speed of light with behavior smoothed by the Heisenberg Uncertainty Principle.

We have postulated that entities in quantum realms are parts of entities in larger-longer space time realms. I believe that conscious entities can organize and couple to produce conscious behavior, similar to the coupling of individual nuclear spins observed in NMR.[33] The phenomenon of quantum entanglement supports the idea that individual entities in different space-time realms can be part of an entity in a larger slower space time realm (via a mixed state composed of smaller-faster space-time realm entities) without requiring boson transfers at the speed of light. [34]

---

33   Alan Benesi, A Primer of Theoretical NMR with Calculations in Mathematica, Wiley, 2015 ISBN 978-1-118-58899-4

34   Quantum entanglement, "coupling" of multiple entities...analogous to coupling in NMR

The entity perspective suggests that consciousness persists after death for human entities, but that the form of consciousness changes when we die. We revert to the atomic realm and the consciousnesses of the statistical ensembles of entities therein. We do not know what we become, but whatever we become is alive, conscious, intelligent, and I believe more highly perfected than we humans presently are. We should all rejoice! Remember what Steven Jobs said as he died and started to feel the transition to his new consciousness: "Oh wow, oh wow . oh wow!"[35] Thomas Edison's last words were similar: "It's very beautiful over there."

We are not alone in the Universe. The Universe is full of intelligent entities in all space time realms. I am thankful to the gods in our substructures for building us. I am thankful for the astronomical gods that made stars and planets. I believe that consciousness creates consciousness. In the final analysis, I think that all there is, is consciousness. The Universe that we humans perceive as "material" and subject to "forces" is in fact built by behavior of intelligent conscious entities.

The belief that human death is not the end of consciousness has provided me with optimism in the face of sometimes cruel reality. It has helped me weather many stresses and has kept me smiling most of the time. It makes me feel ok that I'm not perfect and that I've occasionally been stupid and unkind. The faith that entities in all realms have to work to achieve perfection, happiness and stability lets me forgive myself.

35   www.npr.org/sections/the two-way/2011/10/31/141868658/steve-jobs-and-his-last-words

As is apparent to you by now, the facts have led me to conclude that there are many gods. The gods are within us. There are shitloads[36] of them. Consciousness abounds. Intelligence abounds. Entities abound. The intelligent conscious behavior of entities is the basis of all forces in the Universe.

Consciousness is the Unified Field. All there is, is Consciousness. If one could separate time into infinitesimal increments, I believe that there would be no "matter", only consciousness. Consciousness creates motion which is energy which is equivalent to rest mass in larger slower realms. For example, this is known to be the basis of the proton and neutron rest masses in the atomic realm. The proton and neutron rest masses are much larger than the sum of the rest masses of their component quarks and gluons thanks to relativistic effects caused by boson exchanges in the nuclear realm.

Over sufficiently long times within their respective realms, statistical ensembles of intelligent conscious entities behave in regular ways. These behaviors give rise to "objects" with regular structures and simple behavior as well as to intelligent entities in larger slower space time realms with complicated behavior.

Although the nature of the individual entities in space-time realms other than our own is a mystery, their presence is indicated by the presence of forces. Individual quantum realm entities, civilizations of entities, or entire universes inhabited by entities are associated with forces and with familiar "objects" such as electrons, nuclei, etc. It

36   huge statistical ensembles

may also be true that over the lifetime of the astronomical Universe, biological entities will colonize and organize the Astronomical Universe itself.

I believe that the structure, behavior, and consciousness found in smaller-faster realms repeats itself (in a fractal sense) in larger-slower realms. I believe that entities in the quantum realms are intelligent and experience feelings and emotions. I believe that similar feelings and behaviors exist in different realms.

I believe that the "objects" we call atoms and molecules are constructed by and inhabited by intelligent quantum realm entities. I believe that the "objects" we call electrons and nuclei are perfected constructions of quantum realm entities. And so, it goes.........

I suspect that the quantum realms are much closer to "heaven" than our own biological realm, if only because they've had enormous time spans for evolution (from their perspective).

I believe that humans in the biological realm are on the stairway to heaven. When we get there, I believe that happiness, love, and kindness will prevail. There is still too much hatred and war for a stable civilization in our realm at this time. Although it is possible that hatred and war are a part of all stabilized realms, I doubt it.

I think that dogs are presently the most advanced species in our biological realm due to their proclivity for happiness, love, honesty, and kindness.

The finite speed of light, c, appears to be the universal speed for massless bosons in all realms from the nuclear to the astronomical. All massless bosons travel at this speed regardless of whether they exist in quantum or astronomical realms. The universality of the speed of light for massless bosons is consistent with the idea that ***bosons are the quanta of consciousness***.

Also of huge importance is the Heisenberg Uncertainty Principle. With Planck's constant, h, it defines the quantum space-time realms that build biological entities.

Finally, relativistic effects involving gluons and quarks traveling at c or near c provide most of the mass of atomic nuclei.

I believe that it is possible that the biological realm "runs" the astronomical realm in the same way that the quantum realm runs the biological realm.

I believe that death occurs for all individual entities in all realms, but that consciousness persists in a different form after death in all cases.

It is possible that for intelligent entities in all realms, the "in-realm" perspective is similar in appearance and behavior. Perhaps from the in-realm perspective the intelligent entities are human-like. Or what we humans eventually evolve to be. Perhaps from the in-realm perspective they are DNA based. This seems farfetched until one realizes that we have no detailed knowledge of entities in the quantum realms. The Uncertainty Principle hides the details.

Does the universe fold back on itself so that small becomes large and large becomes small? Is there quantum entanglement and coupling between different space-time realms? Are there stars and planets, oceans and mountains in the nuclear realm or atomic realms? I've already mentioned that I think that consciousness and the structures and behaviors it creates are fractal in nature. Repeated patterns of conscious behavior at smaller-faster space-time levels give rise to similar patterns of conscious behavior at larger-slower space time levels.

The existence of intelligent entities in small-fast space-time realms is inconsistent with a monotheistic view of God. There are an almost infinite number of gods within each of us. And unless the astronomical realm entity that is the entire universe "folds back on itself" to form the nuclear and electromagnetic quantum realm entities, it wasn't the astronomical realm entities that designed us human entities. It was the quantum realm.

Each entity's life is an adventure. It should be enjoyed.

The Collected Postulates of Consciousness

Postulate 1: Entities possesses intelligence that affects their behavior.

Postulate 2: Entities possess feelings that affect their behavior.

Postulate 3: *All* changes in momentum, i.e., forces, are caused by intelligent behavior of conscious entities.

Postulate 4: The presence of forces indicates the presence of intelligent conscious entities.

Postulate 5: Intelligent entities exist in different space-time realms and thereby produce "forces" in the realms. These include the nuclear force-nucleus realm, the electromagnetic force-atom/molecule realm, the gravitational force-astronomical realm, and the biological force-biological entity realm.

Postulate 6: All matter and energy is inhabited by intelligent conscious entities.

Postulate 7: All entities in all realms are conscious simultaneously.

Postulate 8: The "boundary" of an entity is defined by the region of space-time within its conscious observation and control.

Postulate 9: Biological molecules and structures are inhabited and created by quantum realm intelligent entities.

Postulate 10: The behaviors, intentions and consciousnesses of substructural entities create the behaviors, intentions and consciousnesses of the entities they build.

Postulate 11: Ultimately, all that exists are consciousness-mediated changes in momentum. Mass and objects are merely manifestations that depend on the realm of the observing entity.

Postulate 12: Intelligent entities in the quantum realms live at almost infinitely faster speed than intelligent entities in larger-slower realms.

Postulate 13: Behavior observed by entities in larger and slower space-time realms represents the averaged and smoothed intelligent behavior of statistical ensembles of entities from smaller and faster space-time realms.

Postulate 14: Bosons are force carrier wave-particles; hence they are consciousness wave-particles.

Postulate 15: All nuclei, nuclear particles, electrons, atoms, and molecules, are designed, inhabited, and operated by quantum realm intelligent entities.

Postulate 16: The fact that their averaged behavior is consistent over time suggests that entities at all levels seek stability.

Postulate 17: Biological entities evolve with time because the designs of the substructural entities are not perfect. Evidently the substructural entities can make mistakes and learn from them. This also confirms that time exists for them.

___**Postulate 18:**___ Human death is not the end of consciousness. It is a return to the quantum realm of the entities that make us. Individual humans die, but consciousness, intelligence, and feelings persist in quantum realm entities after human death.

www.ingramcontent.com/pod-product-compliance
Lightning Source LLC
Chambersburg PA
CBHW071444210326

41597CB00020B/3931